# First Atlas

## Philip Steele
### Consultant: Clive Carpenter

## Miles Kelly

# Contents

NORTH AMERICA

10

12

14

SOUTH AMERICA

16

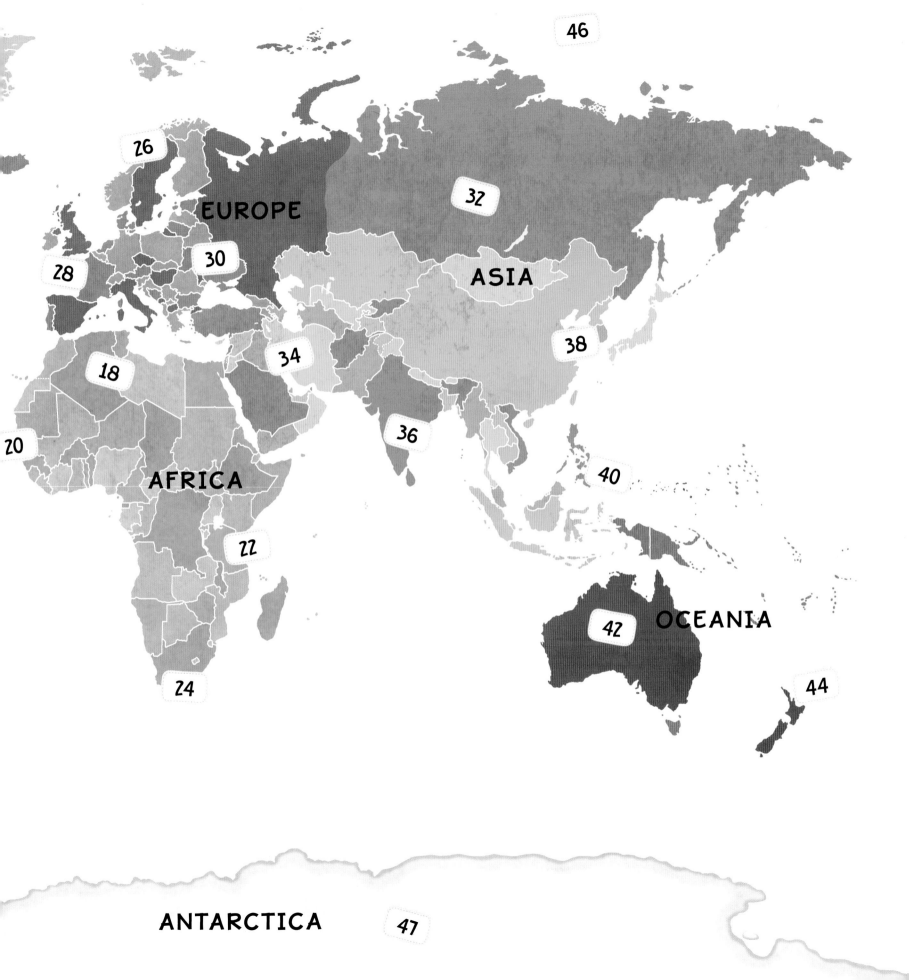

46

26

32

EUROPE

28

30

ASIA

34

18

38

20

36

AFRICA

40

22

42    OCEANIA

24

44

ANTARCTICA    47

# Making maps

We use maps to show important information about our planet, about where we live and how to travel to other places.

The Earth measures 40,030 km around the middle. How can we make a map of something so big? Map-makers use measurements of the Earth's surface, photographs taken from the air and images taken from space by satellites.

To show the Earth on a flat surface, such as a page, we need to show it as a diagram. This means stretching out the Earth's surface, rather like peeling an orange and then squashing the peel flat.

The Earth is a sphere (shaped like a ball), so only a globe can give us a model of the Earth as it really is.

The 'orange peel' map can then be squashed or stretched to create a map without gaps. There are many different ways of doing this, which is why maps don't all look the same.

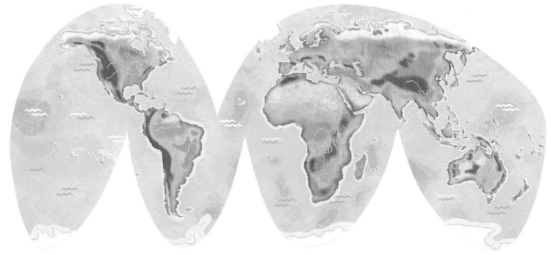

# Using this book

This atlas divides the world into regions. Each region is shown on a map, with countries, places and features marked.

Each map has a locator that highlights where the region is in the world.

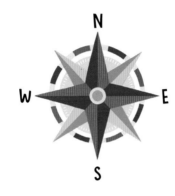

The points of the compass show the direction of north, south, east and west.

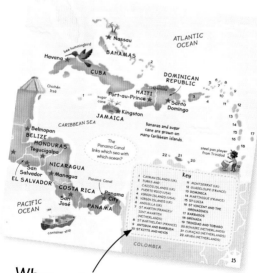

Where places are too small to be labelled on the map, they are listed in a key.

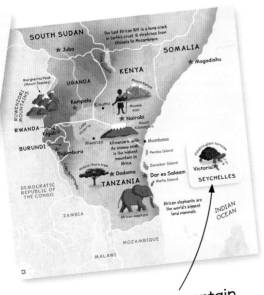

Some places are in a certain region, but are too far away to be included on the map. They are shown in a white box.

## key to maps

Country border

Disputed border – an area that is claimed by more than one country

Capital city – some countries have more than one capital

City

Dependency/territory – an area that is part of or run by another country

Mountains

Rivers

Lakes

5

# Land and sea

Earth's surface is made of a hard layer of rock, called the crust. The thickest parts of the crust form big areas of land called continents. The thinnest parts are covered by the water of the oceans. In total there are seven continents and five oceans.

Earth is home to around 9 million different sorts of plants and animals. They are kept alive by air and water, and warmth and light from the Sun.

ARCTIC OCEAN

EUROPE

ASIA

PACIFIC OCEAN

INDIAN OCEAN

OCEANIA

Salty oceans cover more than two thirds of Earth's surface. Fresh water is found in rivers and lakes.

SOUTHERN OCEAN

ANTARCTICA

**Arctic Circle**
The regions around the North and South Poles are marked by the Arctic and Antarctic Circles.

**Tropic of Cancer**
Map-makers draw lines called the Tropics on either side of the Equator.

**Equator**
Map-makers draw a line called the Equator around the middle of the planet.

**Tropic of Capricorn**

**Antarctic Circle**

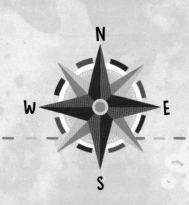

**NORTH AMERICA**

**Vatican City**
Fewer than 900 people live in Vatican City, the world's smallest nation. It is a district of Rome, the capital of Italy.

**AFRICA**

# Countries of the world

**SOUTH AMERICA**

People live in almost every part of the world. They farm the land and build roads, towns and cities. They divide the world up into countries. Each of these countries has its own name, flag, government and capital city. There are nearly 200 countries in the world.

As of 2020, 7.8 billion people live on Earth. About four babies are born every second.

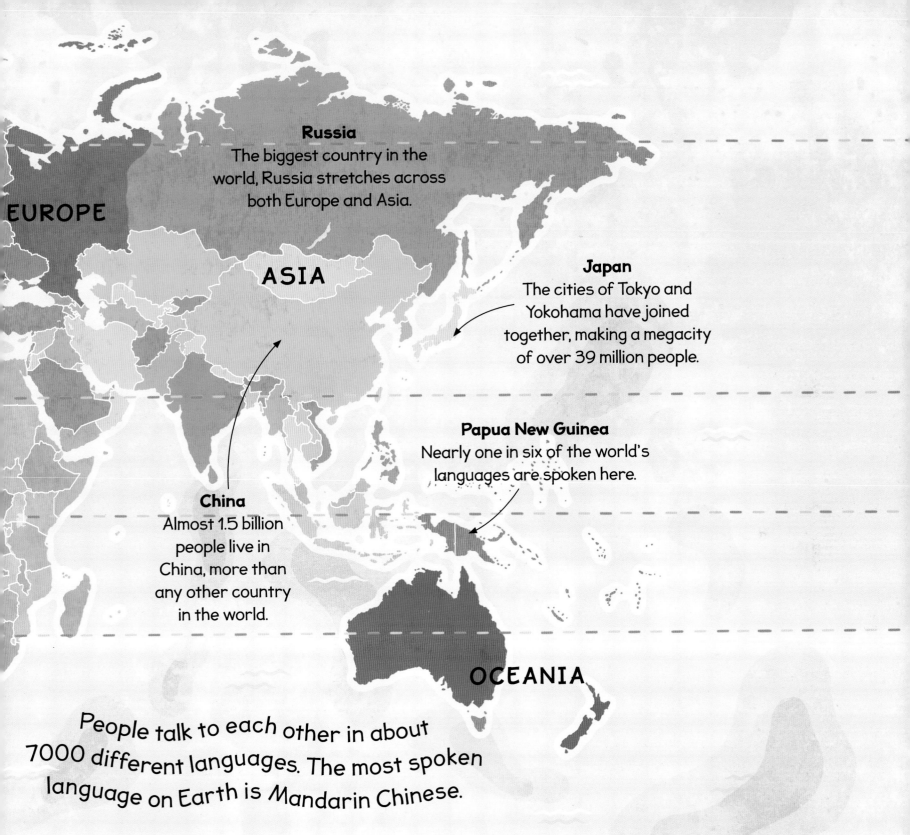

**Russia**
The biggest country in the world, Russia stretches across both Europe and Asia.

EUROPE

ASIA

**Japan**
The cities of Tokyo and Yokohama have joined together, making a megacity of over 39 million people.

**Papua New Guinea**
Nearly one in six of the world's languages are spoken here.

**China**
Almost 1.5 billion people live in China, more than any other country in the world.

OCEANIA

People talk to each other in about 7000 different languages. The most spoken language on Earth is Mandarin Chinese.

ANTARCTICA

# Canada, Alaska and Greenland

The far north of North America is above the Arctic Circle, where it is very cold and the ocean freezes over. Further south, it is a bit warmer. There are forests, lakes and grassy plains called prairies, with huge fields of wheat. In the west are high mountains.
Most people live in cities in the south.

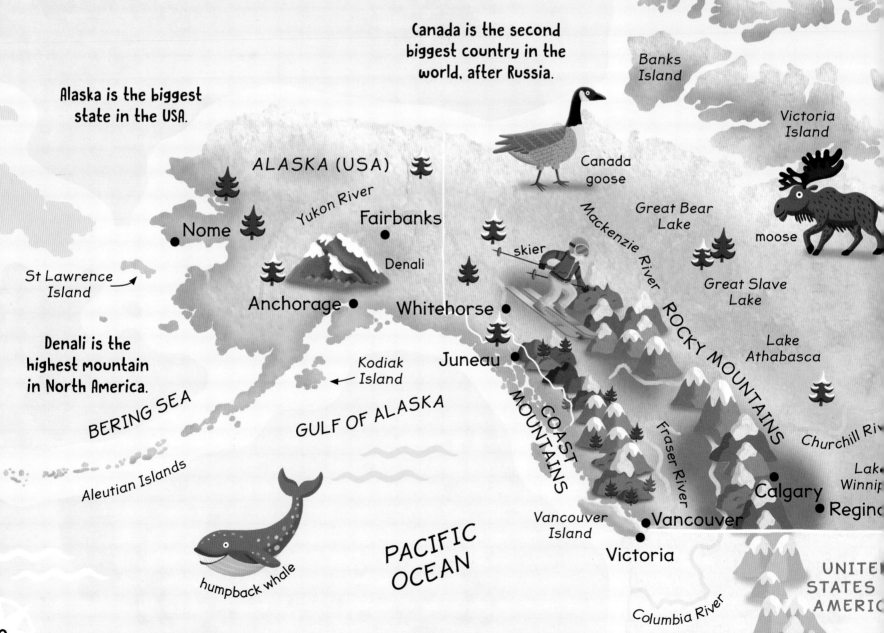

Canada is the second biggest country in the world, after Russia.

Alaska is the biggest state in the USA.

Banks Island

Victoria Island

ALASKA (USA)

Canada goose

Nome

Yukon River

Fairbanks

Great Bear Lake

moose

Denali

skier

St Lawrence Island

Anchorage

Whitehorse

Mackenzie River

Great Slave Lake

ROCKY MOUNTAINS

Denali is the highest mountain in North America.

Juneau

Kodiak Island

Lake Athabasca

BERING SEA

GULF OF ALASKA

COAST MOUNTAINS

Churchill Riv

Aleutian Islands

Fraser River

Lake Winnip

Calgary

Regina

Vancouver Island

Vancouver

humpback whale

PACIFIC OCEAN

Victoria

UNITE STATES AMERIC

Columbia River

ARCTIC OCEAN

glacier

Ellesmere
Island

polar bear

BAFFIN
BAY

Baffin Island

Arctic tern

GREENLAND
(DENMARK)

snowmobile

Greenland is the world's
biggest island. It is mostly
covered by a massive
sheet of ice.

ANADA

Nuuk
(Godthåb)

ICELAND

HUDSON BAY

Churchill

ATLANTIC
OCEAN

Which long
water system links the
Great Lakes with the
Atlantic Ocean?

ice hockey
player

red maple
tree

St Lawrence River
and Seaway

innipeg

CN Tower

Newfoundland

St John's

Great Lakes

Québec

Ottawa

Montréal

Nova
Scotia

Toronto

Halifax

N

W          E

S

# United States of America

The USA is a large country, made up of 50 states. Most of them lie between the Atlantic and the Pacific Oceans, except for Alaska (see page 10) and Hawaii (see page 44). The USA has many big, busy cities. It also has peaceful forests and lakes, and grassy plains called prairies. There are high mountains, hot deserts and ocean shores too.

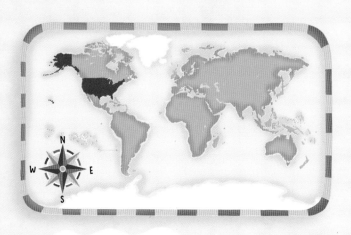

Seattle

sequoia tree

Columbia River

CASCADE RANGE

Over millions of years the Colorado River carved out a huge gorge, forming the Grand Canyon.

Golden Gate Bridge

SIERRA NEVADA MOUNTAINS

San Francisco

Colorado River

Grand Canyon

Los Angeles

cactus

PACIFIC OCEAN

N
W
E
S

The mighty Rocky Mountains run down North America for 4800 km.

CANADA

The Great Lakes contain over one fifth of all the fresh water on Earth.

bald eagle

Lake Superior

Lake Huron

Lake Ontario

Statue of Liberty

Lake Michigan

Missouri River

Detroit

Lake Erie

Niagara Falls

Boston

New York City

White House

UNITED STATES OF AMERICA

Chicago

St Louis

APPALACHIAN MOUNTAINS

Washington DC

tornado

The USA's two longest rivers, the Missouri and the Mississippi, join together near the city of St Louis.

Arkansas River

Mississippi River

ATLANTIC OCEAN

Atlanta

American alligator

Dallas

raccoon

New Orleans

Houston

MEXICO

GULF OF MEXICO

Washington DC, the USA's capital, is not in any state but in its own special area, the District of Columbia (DC).

Miami

13

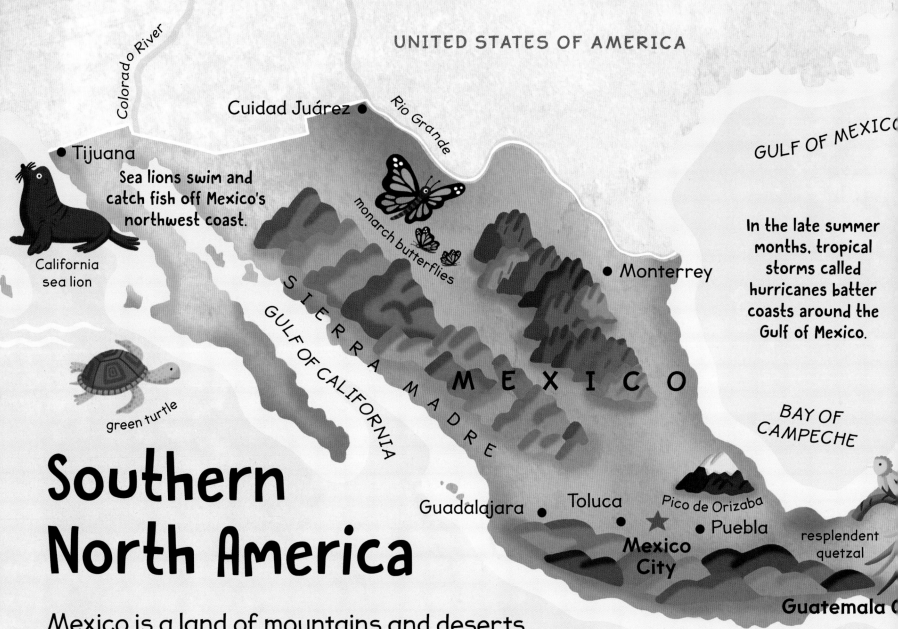

Colorado River

Cuidad Juárez

Rio Grande

GULF OF MEXICO

Tijuana

Sea lions swim and catch fish off Mexico's northwest coast.

monarch butterflies

California sea lion

Monterrey

In the late summer months, tropical storms called hurricanes batter coasts around the Gulf of Mexico.

SIERRA MADRE

GULF OF CALIFORNIA

MEXICO

green turtle

BAY OF CAMPECHE

Guadalajara

Toluca

Pico de Orizaba

Puebla

Mexico City

resplendent quetzal

# Southern North America

Guatemala C

GUATEMALA

Mexico is a land of mountains and deserts, volcanoes and tropical forests, and ruins of ancient cities. Over 21 million people live in or around the capital, Mexico City. The smaller countries of Central America narrow to a thin strip of land, which is crossed by the Panama Canal. To the east are chains of tropical islands.

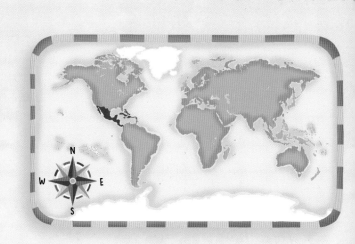

ATLANTIC OCEAN

bee hummingbird

Nassau

BAHAMAS

2

DOMINICAN REPUBLIC

5  6  8
  7
    9
  10

Havana

CUBA

Chichén Itzá

Hispaniola

HAITI
Port-au-Prince

Santo Domingo

3

4  11  12

13

14

15  17

16

18

1

sugar cane

Kingston

JAMAICA

Bananas and sugar cane are grown on many Caribbean islands.

steel pan player from Trinidad

19

CARIBBEAN SEA

Belmopan

BELIZE

HONDURAS

Tegucigalpa

The Panama Canal links which sea with which ocean?

22  21  20

NICARAGUA

San Salvador

SALVADOR

Managua

Panama Canal

COSTA RICA

San José

PANAMA

Panama City

PACIFIC OCEAN

container ship

COLOMBIA

## key

| | | |
|---|---|---|
| 1 | CAYMAN ISLANDS (UK) | 11 MONTSERRAT (UK) |
| 2 | TURKS AND CAICOS ISLANDS (UK) | 12 GUADELOUPE (FRANCE) |
| 3 | PUERTO RICO (USA) | 13 DOMINICA |
| 4 | VIRGIN ISLANDS (USA) | 14 MARTINIQUE (FRANCE) |
| 5 | VIRGIN ISLANDS (UK) | 15 ST LUCIA |
| 6 | ANGUILLA (UK) | 16 ST VINCENT AND THE GRENADINES |
| 7 | ST MARTIN (FRANCE)/ SINT MAARTEN (NETHERLANDS) | 17 BARBADOS |
| 8 | ST BARTHELÉMY (FRANCE) | 18 GRENADA |
| 9 | ANTIGUA AND BARBUDA | 19 TRINIDAD AND TOBAGO |
| 10 | ST KITTS AND NEVIS | 20 BONAIRE (NETHERLANDS) |
| | | 21 CURAÇAO (NETHERLANDS) |
| | | 22 ARUBA (NETHERLANDS) |

# South America

The continent of South America stretches all the way from the warm waters of the Caribbean Sea to the cold winds and waves of Cape Horn.

The world's largest tropical rainforest surrounds the mighty Amazon River. There are also deserts and grasslands, farms and bustling cities.

FRENCH GUIANA (FRANCE)

Cayenne

Georgetown

Paramaribo

GUYANA

SURINAME

VENEZUELA

Caracas

Maracaibo

Lake Maracaibo

Orinoco River

Angel Falls

Cartagena

PANAMA

Medellín

Bogotá

COLOMBIA

Cali

Quito

ECUADOR

Guayaquil

PERU

Lima

Cusco

Machu Picchu

ANDES MOUNTAINS

Belém

Fortaleza

Recife

Salvador

golden lion tamarin

scarlet macaw

São Francisco River

Tocantins River

Manaus

Amazon River

A M A Z O N   R A I N F O R E S T

The Amazon is a gigantic river, about 6400 km long.

coffee beans

anaconda

BRAZIL

Brasília

BOLIVIA

La Paz

Lake Titicaca

Bolivian

P A
W

N   E
W   S

16

Rio de Janeiro

São Paulo

Curitiba

Christ the
Redeemer
statue

bottlenose dolphins

Porto Alegre

ATLANTIC
OCEAN

There may be
390 billion trees in the
Amazon rainforest. More
types of plants and animals
are found here than
anywhere else on Earth.
There are huge snakes,
noisy monkeys, colourful
tree frogs and bright-
green parrots.

SOUTH
GEORGIA
(UK)

PARAGUAY

Asunción

Montevideo

Río de la Plata

Parsná River

URUGUAY

Buenos
Aires

Bahía Blanca

The Andes Mountains cross
seven countries. They form
the longest mountain
range in the world.

Southern
rockhopper
penguin

S

Andean condor

ARGENTINA

Córdoba

Rosario

PAMPAS GRASSLANDS

FALKLAND
ISLANDS (UK)

CHILE

ATACAMA DES

Valparaíso
Santiago

Mount
Aconcagua

Concepción

A N D E S   M O U N T A I N S

P A T A G O N I A

Ushuaia

Cape Horn

It hardly ever
rains in the Atacama
Desert. It is one of
the driest places
on the planet.

Tierra del Fuego

PACIFIC
OCEAN

# Northern Africa

Most of Northern Africa is covered by the Sahara, the world's biggest hot desert. The Nile River flows through Sudan and Egypt to the Mediterranean Sea. Along the coast, farmers pick olives or oranges, and people shop at busy markets. South of the Sahara are dry lands, where people struggle to grow crops or keep cattle or goats.

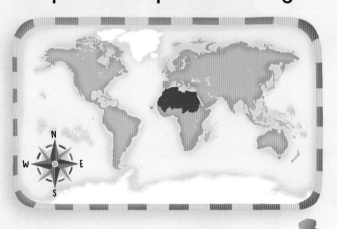

The Pyramids of Giza were built over 4500 years ago. They were tombs for the rulers of ancient Egypt.

The Great Mosque of Djenné, built from mud, is the most famous landmark in Mali.

Ceuta (SPAIN)
Tangier
Me (SF

Fes

**MOROCCO**

CANARY ISLANDS (SPAIN)

ATLAS MOU

solar power plant

★ El Aaiún

**WESTERN SAHARA**

Morocco has lots of sunshine. It uses it to make electricity — this is called solar power.

**MA**

ATLANTIC OCEAN

**MAURITANIA**

★ Nouakchott

Great Mosque of Djenné

**CAPE VERDE**

★ Praia

SENEGAL

Bamako ★

GUINEA

BUR F

IVORY COAST

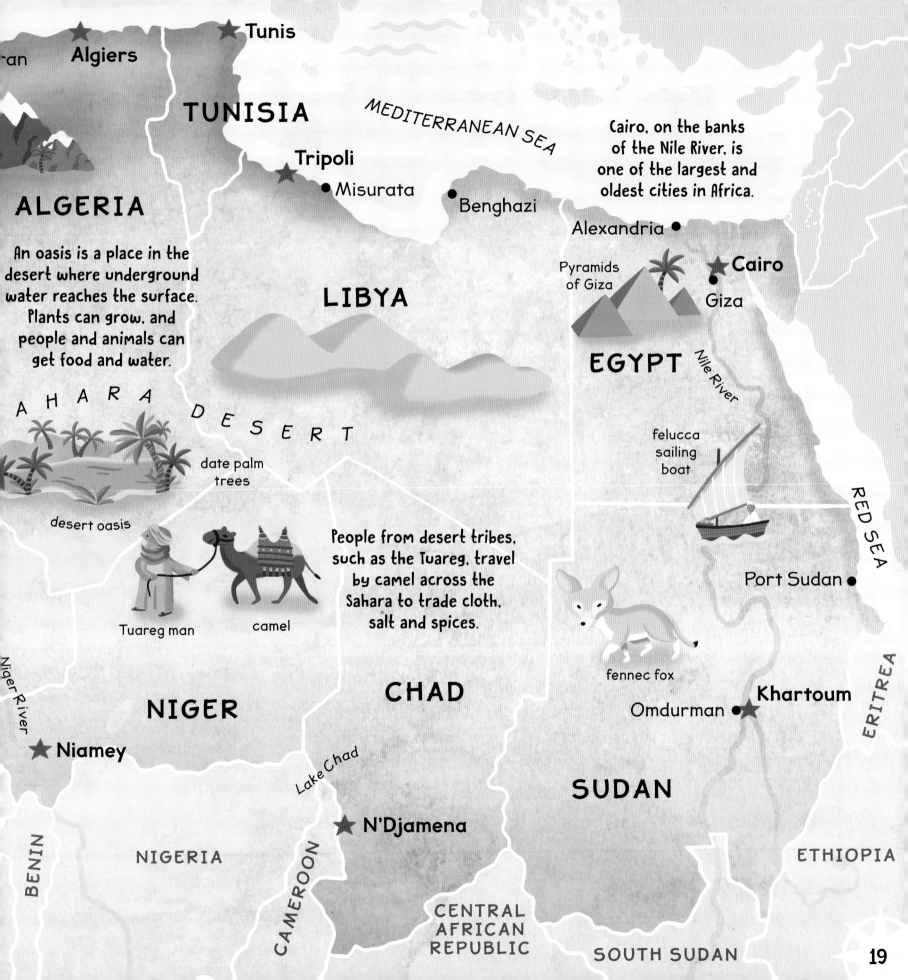

ran

Algiers

★ Tunis

TUNISIA

MEDITERRANEAN SEA

Cairo, on the banks
of the Nile River, is
one of the largest and
oldest cities in Africa.

**ALGERIA**

★ Tripoli
• Misurata
• Benghazi

Alexandria •

An oasis is a place in the
desert where underground
water reaches the surface.
Plants can grow, and
people and animals can
get food and water.

**LIBYA**

Pyramids
of Giza

★ **Cairo**

• Giza

**EGYPT**

Nile River

A H A R A   D E S E R T

felucca
sailing
boat

date palm
trees

desert oasis

People from desert tribes,
such as the Tuareg, travel
by camel across the
Sahara to trade cloth,
salt and spices.

RED SEA

Port Sudan •

Tuareg man          camel

fennec fox

**CHAD**

Khartoum

Omdurman •  ★

Niger River

**NIGER**

★ Niamey

Lake Chad

**SUDAN**

ERITREA

BENIN

NIGERIA

★ N'Djamena

CAMEROON

CENTRAL
AFRICAN
REPUBLIC

SOUTH SUDAN

ETHIOPIA

MAURITANIA

ATLANTIC OCEAN

Senegal River

MALI

BURKINA FASO

Ouagadougou ★

Niger River

Dakar ★ SENEGAL

Banjul ★
GAMBIA

Bissau ★
GUINEA-BISSAU

Conakry ★

Freetown ★

SIERRA LEONE

GUINEA

BENIN

TOGO

GHANA

Ibad●

IVORY COAST

cocoa beans

Lake Volta

Lagȯ●

Porto Nova

Lomé

Accra ★

LIBERIA

Yamoussoukro ★

Abidjan ●

Monrovia ●

GULF OF GUINEA

The village of Debundscha, near Mount Cameroon, is the rainiest place in Africa.

fishing boat

# West and Central Africa

It is hot and very humid along the West African coast. Oil palms and cocoa grow well in this climate. There are huge crowded cities such as Lagos in Nigeria. The north is drier, with plains where winds carry dust from the desert. Central Africa lies on the Equator. Beyond the cities of Brazzaville and Kinshasa, the Congo River winds through thick rainforest.

SUDAN

Around 200 million
people live in Nigeria,
more than anywhere
else in Africa.

CHAD

SOUTH
SUDAN

ano ●         yams

Gorillas are the biggest
and most powerful apes.
They live in the forests
of Central Africa.

NIGERIA

CENTRAL
AFRICAN
REPUBLIC

Abuja
★

*Benue River*

CAMEROON

The Congo River is
4370 km long. It is the
world's deepest river.

Bangui ★

*Congo River*

UGANDA

mountain gorilla

● Kisangani

Mount
Cameroon

Port
Harcourt
●

Yaoundé
★

Douala ●

Malabo ★

*Bioko Island*

Djibloho ★  EQUATORIAL
GUINEA

okapi

C O N G O   R A I N F O R E S T

RWANDA

BURUNDI

O TOMÉ
AND
ÍNCIPE

São
Tomé
★

Libreville
★

CONGO

GABON

DEMOCRATIC
REPUBLIC OF
THE CONGO

TANZANIA

Brazzaville
★

Kinshasa
★

Mbuji-Mayi ●

Nile
crocodile

CABINDA
(ANGOLA)

Lubumbashi
●

Can you find on the
map three countries with
the word 'Guinea' in the
name, and two with
'Congo'?

ANGOLA

ZAMBIA

# East Africa

Lions roar while herds of elephants, zebras and giraffes cross the savannah, grassy plains dotted with thorn trees and rocks. In the west are great lakes that fill the deep valleys of the East African Rift. To the east are the white sands, blue waters and small islands of the Indian Ocean. There are also big ports at the cities of Mombasa and Dar es Salaam.

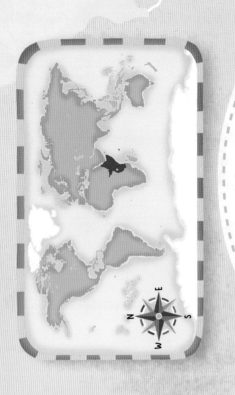

## Learn to speak Swahili

hello *habari* (hab-ah-ree)

goodbye *kwaheri* (kwah-he-ree)

thank you *asante* (ah-sahn-teh)

Swahili is spoken and understood by many peoples in East Africa, especially in Kenya and Tanzania.

Over 800 years ago, churches like this were carved out of rock in northern Ethiopia.

rock church

GULF OF ADEN

SOMALILAND

DJIBOUTI
★ Djibouti

ETHIOPIA

RED SEA

★ Asmara

ERITREA

HORN OF AFRICA

zebra

Blue Nile River

SUDAN

White Nile River

the Sudd swamp

In South Sudan the White Nile River flows through an enormous swamp called the Sudd.

N
E
S
W

The East African Rift is a long crack in Earth's crust. It stretches from Ethiopia to Mozambique.

SOMALIA

Mogadishu

SEYCHELLES

Victoria

Aldabra giant tortoise

INDIAN OCEAN

KENYA

Mount Kenya

Maasai man

Nairobi

Mount Kilimanjaro

Mombasa

Pemba Island

Zanzibar Island

Dar es Salaam

Mafia Island

African elephants are the world's biggest land mammals.

Lake Turkana

Kisumu

Juba

UGANDA

Kampala

Lake Victoria

Mwanza

Kilimanjaro, with its snowy peak, is the highest mountain in Africa.

Dodoma

TANZANIA

acacia thorn tree

African elephant

MOZAMBIQUE

Margherita Peak (Mount Stanley)

RUWENZORI MOUNTAINS

RWANDA

Kigali

BURUNDI

Bujumbura

Lake Tanganyika

DEMOCRATIC REPUBLIC OF THE CONGO

ZAMBIA

MALAWI

23

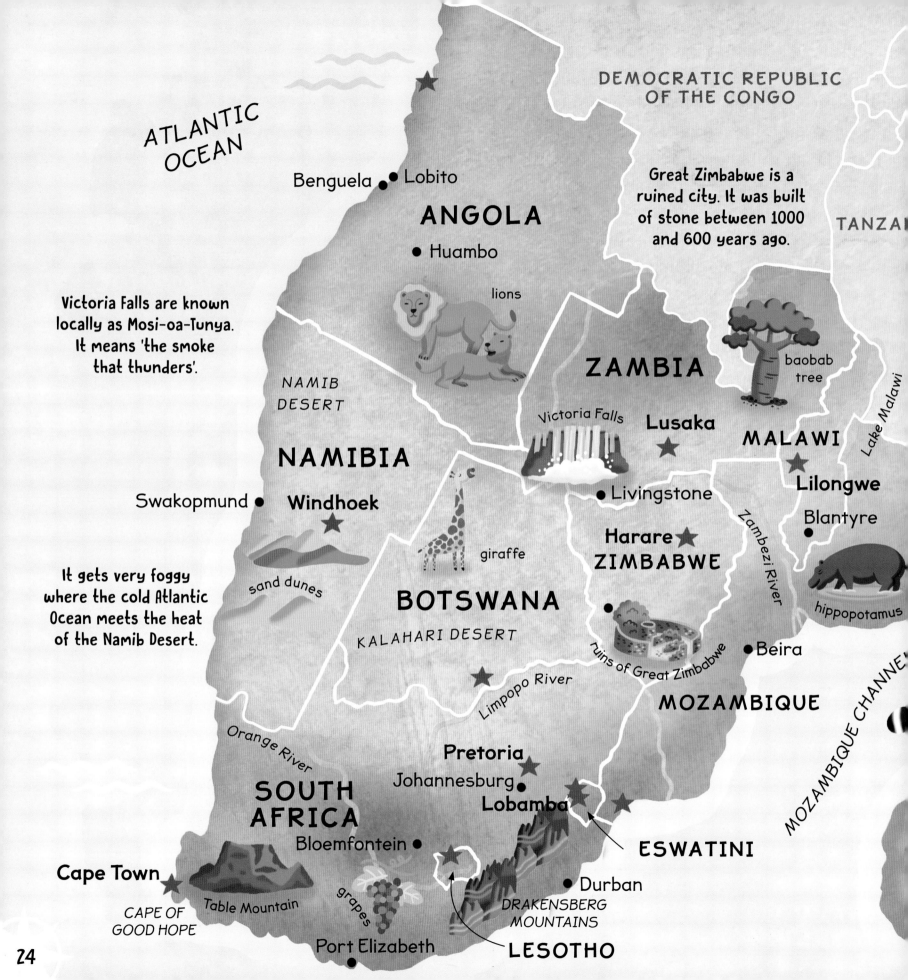

ATLANTIC
OCEAN

DEMOCRATIC REPUBLIC
OF THE CONGO

TANZA

Benguela • Lobito

ANGOLA

• Huambo

lions

Great Zimbabwe is a
ruined city. It was built
of stone between 1000
and 600 years ago.

baobab
tree

ZAMBIA

Victoria Falls are known
locally as Mosi-oa-Tunya.
It means 'the smoke
that thunders'.

NAMIB
DESERT

Victoria Falls

Lusaka

MALAWI

NAMIBIA

Lilongwe

Swakopmund •   Windhoek

• Livingstone

Blantyre

giraffe

Harare

It gets very foggy
where the cold Atlantic
Ocean meets the heat
of the Namib Desert.

sand dunes

ZIMBABWE

Zambezi River

BOTSWANA

hippopotamus

KALAHARI DESERT

ruins of Great Zimbabwe

• Beira

Limpopo River

MOZAMBIQUE

Orange River

Pretoria

Johannesburg •

SOUTH
AFRICA

Lobamba

ESWATINI

MOZAMBIQUE CHANNEL

Bloemfontein •

Cape Town

• Durban

CAPE OF
GOOD HOPE

Table Mountain

grapes

DRAKENSBERG
MOUNTAINS

Port Elizabeth

LESOTHO

24

Many of the wild animals on the island of Madagascar are found nowhere else on Earth. They include furry lemurs, colourful lizards and hundreds of types of frogs and birds.

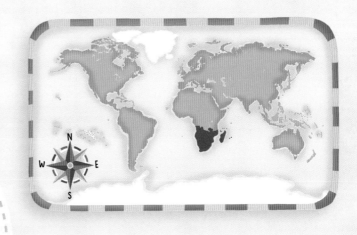

INDIAN OCEAN

South Africa's Mponeng gold mine is the world's deepest mine. It reaches up to 4 km underground.

COMOROS
★ Moroni

MAYOTTE
(FRANCE)

MADAGASCAR

★ Antananarivo

-tailed lemur

MAURITIUS
★ Port Louis

RÉUNION
(FRANCE)

# Southern Africa

The waters of Lake Malawi can be peaceful and still, but the Zambezi River runs wild as it roars over Victoria Falls. Many of the grasslands and farms of Southern Africa suffer if the rains don't come. The driest parts are the Kalahari and the Namib Deserts. In South Africa there are diamond and gold mines, and grapes, apples and peaches are grown in the mild climate of the far south.

# Northern Europe

Europe's far north reaches the icy Arctic Ocean. Iceland is an island with volcanoes, hot springs and geysers that shoot water and steam into the air. Finland has around 188,000 lakes. Norway, Sweden and Denmark are known as the Scandinavian nations. You can cross the Baltic Sea to visit Estonia, Latvia and Lithuania.

In the snowy north of Europe, the Sami people herd reindeer.

ARCTIC OCEAN

BARENTS SEA

Northern Lights

SVALBARD (NORWAY)

GREENLAND SEA

Near the North Pole, the sun doesn't set in the middle of summer, so it stays light all night long. In the middle of winter the sun doesn't rise, so it is dark all day. When it is winter in the Arctic, it is summer in the Antarctic.

## ICELAND

**Reykjavik** Strokkur geyser

FAROE ISLANDS (DENMARK)

**Tórshavn**

reindeer

Sami woman

North Cape

● Tromsø

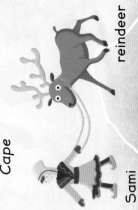

Deep, narrow sea inlets are called fjords.

Large parts of Finland are covered in forest.

The Baltic Sea ports have been centres of trade for hundreds of years.

**FINLAND**
- Oulu
- Tampere
- Helsinki

**SWEDEN**
- Turku
- Åland Islands
- Uppsala
- Västerås
- **Stockholm**
- Linköping
- Örebro
- Göteborg
- Gotland
- Bornholm

GULF OF BOTHNIA

GULF OF FINLAND

BALTIC SEA

Lake Peipus

**ESTONIA**
- Tallinn

**LATVIA**
- Riga

**LITHUANIA**
- Kaunas
- **Vilnius**

Trakai Island Castle

RUSSIA

BELARUS

KALININGRAD (RUSSIA)

POLAND

GERMANY

**NORWAY**
- Bergen
- Trondheim
- **Oslo**
- Stavanger

NORWEGIAN SEA

SCANDINAVIAN MOUNTAINS

Dal River

Klaralven River

Glomma River

Lake Vänern

Lake Vättern

SKAGERRAK

NORTH SEA

**DENMARK**
- Århus
- **Copenhagen**
- Malmö

Little Mermaid statue

27

# Western Europe

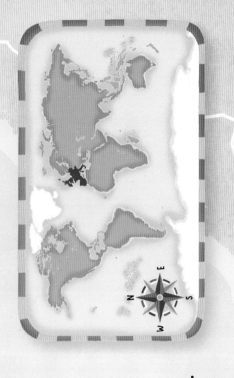

A warm current in the Atlantic Ocean gives Northwestern Europe a mild and rainy climate, making hills, woods and fields green. Southwestern Europe borders the Mediterranean Sea. It has hot summers that bring many tourists to the beaches. The high peaks of the Alps and Pyrenees are covered in snow. Across Western Europe there are lots of historical castles, churches and villages, but also gleaming modern cities.

The Netherlands is low-lying and very flat. Large parts are below sea level, so it has to be protected from flooding by huge sea walls and pumps.

The Rhine River flows from Switzerland to the North Sea, passing through Germany and the Netherlands.

BALTIC SEA

DENMARK

Hamburg

NORTH SEA

NETHERLANDS

Shetland Islands

Orkney Islands

Highland cattle

puffin

SCOTLAND

Glasgow • Edinburgh

UNITED KINGDOM

Leeds

Giant's Causeway

NORTHERN IRELAND • Belfast

IRISH

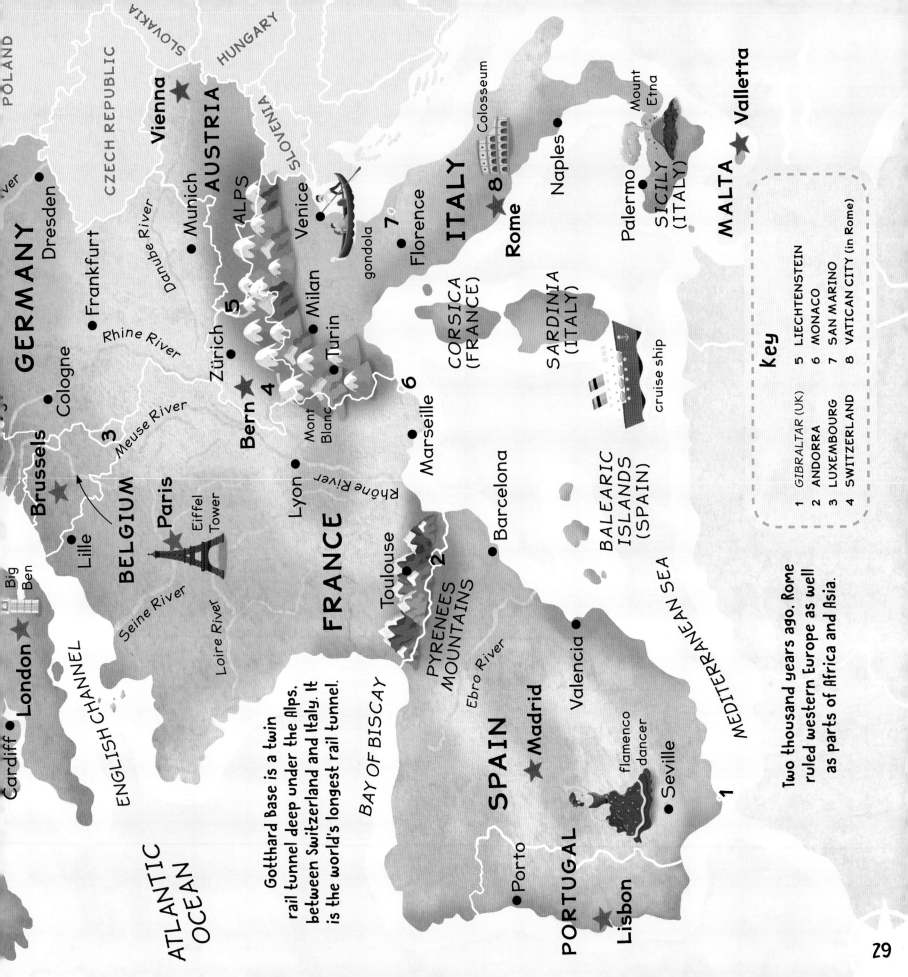

POLAND

SLOVAKIA

CZECH REPUBLIC

HUNGARY

GERMANY

...ver
Dresden
Frankfurt
Cologne
Brussels
BELGIUM

Munich
Vienna ★
AUSTRIA
SLOVENIA
ALPS
Zürich
Bern ★
4
5
Venice
Milan
Turin
Mont Blanc

Rhine River
Danube River
Meuse River

Lille
Paris ★
Eiffel Tower
Seine River
Loire River

FRANCE

Lyon
Rhône River
Toulouse
2
PYRENEES MOUNTAINS

Marseille
6

CORSICA (FRANCE)

SARDINIA (ITALY)

cruise ship

gondola
7
Florence

ITALY
8 ★
Rome

Colosseum

Naples

Palermo

Mount Etna

SICILY (ITALY)

MALTA ★ Valletta

key

1 GIBRALTAR (UK)
2 ANDORRA
3 LUXEMBOURG
4 SWITZERLAND

5 LIECHTENSTEIN
6 MONACO
7 SAN MARINO
8 VATICAN CITY (in Rome)

Two thousand years ago, Rome ruled western Europe as well as parts of Africa and Asia.

Barcelona

BALEARIC ISLANDS (SPAIN)

MEDITERRANEAN SEA

Valencia

Madrid ★
SPAIN

Seville

flamenco dancer

1

Porto

PORTUGAL

Lisbon ★

Cardiff
London ★
Big Ben

ENGLISH CHANNEL

ATLANTIC OCEAN

Gotthard Base is a twin rail tunnel deep under the Alps, between Switzerland and Italy. It is the world's longest rail tunnel.

BAY OF BISCAY

Ebro River

3

29

LITHUANIA

BELARU

GERMANY

Gdansk

European bison

**POLAND**

⭐ Warsaw

decorated Easter eggs

SUDETEN MOUNTAINS

Oder River

Katowice

Kraków

UKR

**Can you find on the map:**

the capital of the Czech Republic?

five countries that share a border with Romania?

the continent that borders Eastern Europe?

Prague ⭐

**CZECH REPUBLIC**

TATRA MOUNTAINS

2

⭐ Bratislava

AUSTRIA

Lake Balaton

CARPATHIAN MOUNTAINS

1 Chisina

⭐ Budapest

**HUNGARY**

Ljubljana

⭐ 3

The Danube River winds through ten countries before reaching the Black Sea. It flows through the centre of Budapest, capital of Hungary.

⭐ **CROATIA**

Zagreb

**ROMANIA**

Danube River

4

Belgrade ⭐

⭐ Buchar

ADRIATIC SEA

Sarajevo ⭐

6

Pristina

BALKAN MOUNTAINS

ITALY

5

7

⭐ Podgorica

8

9

⭐ Sofia

10

Tirana ⭐

⭐ Skopje

● Thessaloníki

**Key**

1 MOLDOVA
2 SLOVAKIA
3 SLOVENIA
4 BOSNIA-HERZEGOVINA
5 MONTENEGRO
6 SERBIA
7 KOSOVO
8 ALBANIA
9 NORTH MACEDONIA
10 BULGARIA

The ancient Greek temple of the Parthenon still tops the skyline in Athens. It is more than 2400 years old.

**GREECE**

Parthenon temple

Olympia ●

⭐ **Athens**

AEGEAN SEA

Crete

great hammerhead shark

Many people in Central and Eastern Europe paint eggs with finely detailed patterns, for the Christian festival of Easter.

# Central and Eastern Europe

A great plain stretches across Poland, where there are lakes, farmland, ancient forests and industrial cities. Hills and mountains rise to the south, through the Czech Republic, Slovakia, Romania and Bulgaria. There are many historical cities in Central Europe. In the south is Greece, with its islands and mountains rising from deep-blue seas.

ev

Kharkov ●

Dnieper River

● Dnipropetrovsk

Donetsk ●

great white pelican

essa

CRIMEA

BLACK SEA

RUSSIA

TURKEY

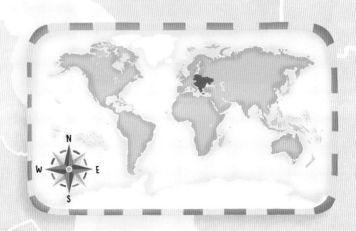

The far west of Turkey is in Europe, but most of the country is in Asia.

rry

Mount Elbrus is the highest mountain in Europe, towering 5642 m above sea level. It is part of the Caucasus Mountains.

*Severnaya Zemlya*

ARCTIC OCEAN

*Novaya Zemlya*

Russia has the largest number of Eurasian brown bears in the world.

FINLAND

*Eurasian brown bear*

ballet dancer

● St Petersburg

Nizhny Novgorod

U R A L   M O U N T A I N S

*Irtysh River*

*Yenisei River*

*Ob River*

R U S S I A

*Angara River*

St Basil's Cathedral

★ **Moscow**

● Yekaterinburg

● Novosibirsk

★ **Minsk**

*Don River*

*Volga River*

**BELARUS**

★ **Astana**

SAYAN MOUNTAINS

Lak

UKRAINE

*Ural River*

space launch

**KAZAKHSTAN**

MONGOLI

Volgograd ●

CAUCASUS MOUNTAINS

Mount Elbrus

*Aral Sea*

● Almaty

★ **Bishkek**

**Key**

BLACK SEA

CASPIAN SEA

1

★ **Tashkent**

7

1 GEORGIA

2 ARMENIA

**Tbilisi**

2  3

4

5

3 AZERBAIJAN

4 TURKMENISTAN

TURKEY

**Yerevan**

3

★ **Ashgabat**

★ **Dushanbe**

6

5 UZBEKISTAN

6 TAJIKISTAN

**Baku**

AFGHANISTAN

7 KYRGYZSTAN

IRAN

icebreaker ship

The Trans-Siberian Railway links Moscow with Russia's Pacific coast, a journey of 9289 km that takes just over six days. It is the world's longest railway line.

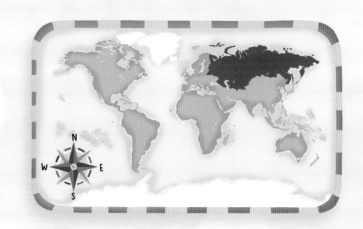

Lena River

VERKHOYANSK MOUNTAINS

KOLYMA MOUNTAINS

S I B E R I A

Trans-Siberian Railway

Amur River

CHINA

SEA OF OKHOTSK

Kamchatka volcanoes

● Vladivostok

The Baikonur Cosmodrome in Kazakhstan is the world's oldest and biggest space launch site. It is used to launch astronauts to the International Space Station.

# Russia and its neighbours

Russia's two biggest cities, Moscow and St Petersburg, are in Europe. East of the Ural Mountains are the vast open spaces of Siberia, in Northern Asia. Frozen plains known as tundra border the Arctic Ocean. Great forests cover much of Russia. To the south are the mountains, grasslands and deserts of Central Asia.

# Southwest Asia

Vast deserts of sand and rock shimmer in the heat. In the cooler, greener regions there are grassy hills, high mountains and rolling rivers such as the Euphrates and the Tigris. Where there is water, farmers grow olives, wheat, figs and dates, or raise sheep and goats. There are ancient cities, but also modern skyscrapers and oil wells.

Asia meets Europe at Istanbul, in Turkey.

AFGHANISTAN

TURKMENISTAN

IRAN

Mashhad

ELBURZ MOUNTAINS

Mount Damavand

Tehran

Esfahan

ruins of the ancient city of Persepolis

ZAGROS MOUNTAINS

CASPIAN SEA

AZERBAIJAN

ARMENIA

GEORGIA

Tabriz

Lake Urmia

Mosul

Kirkuk

Baghdad

Tigris River

Euphrates River

Basra

IRAQ

Mount Ararat

Lake Van

olive tree

Aleppo

SYRIA

Damascus

Amman

3

Beirut

1

Jerusalem

2

Dead Sea

WEST BANK

GAZA STRIP

MEDITERRANEAN SEA

CYPRUS

Nicosia

lemons

BLACK SEA

PONTIC MOUNTAINS

TURKEY

Ankara

Konya

Istanbul

Bursa

Izmir

Antalya

GULF OF OMAN

Muscat ★

Arabian oryx (antelope)

ARABIAN SEA

Dubai ●

8

Abu Dhabi ★

7

OMAN

The Burj Khalifa skyscraper in Dubai is the world's highest building, at 829.8 m.

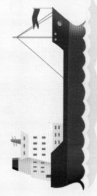

oil tanker

SOCOTRA (YEMEN)

...GULF

Manama ★

5

Doha ★

6

Riyadh ★

SAUDI ARABIA

common kestrel

A R A B I A N   D E S E R T

YEMEN

San'a ★

GULF OF ADEN

● Aden

Thousands of years ago, people in this part of the world...

• built the first cities
• were the first farmers
• invented ways of writing
• invented the wheel

Kaaba shrine

Medina ●

● Mecca

Jeddah ●

RED SEA

Jerusalem is holy to Jews, Muslims and Christians.

Every year millions of Muslim pilgrims travel from all over the world to the holy city of Mecca. The Kaaba shrine is the holiest place in Islam.

key

1 LEBANON
2 ISRAEL
3 JORDAN
4 KUWAIT
5 BAHRAIN
6 QATAR
7 UNITED ARAB EMIRATES
8 PART OF OMAN

The Taj Mahal at Agra in India is a beautiful marble tomb, built over 360 years ago for the wife of the emperor.

Kabul ★

• Herat

**AFGHANISTAN**

Islama

IRAN

• Kandahar

La

**PAKISTAN**

Indus River

greater flamingo

TH

• Karachi

Ahmadaba

ARABIAN SEA

Sur

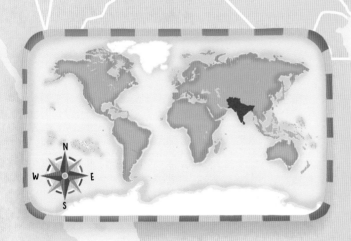

N
W E
S

Mumb

# Southern Asia

Many of the world's highest mountains rise in Pakistan, Nepal, Bhutan and China. Melting snows feed great rivers such as the Indus and the Ganges, which cross hot, dusty plains to the south. Each summer, winds called monsoons bring heavy rains to the dry fields. Over 1.3 billion people live in India, speaking many different languages. India narrows to a southern point towards the island of Sri Lanka.

INDIAN OCEAN

**MALDIVES**
**Malé**
★

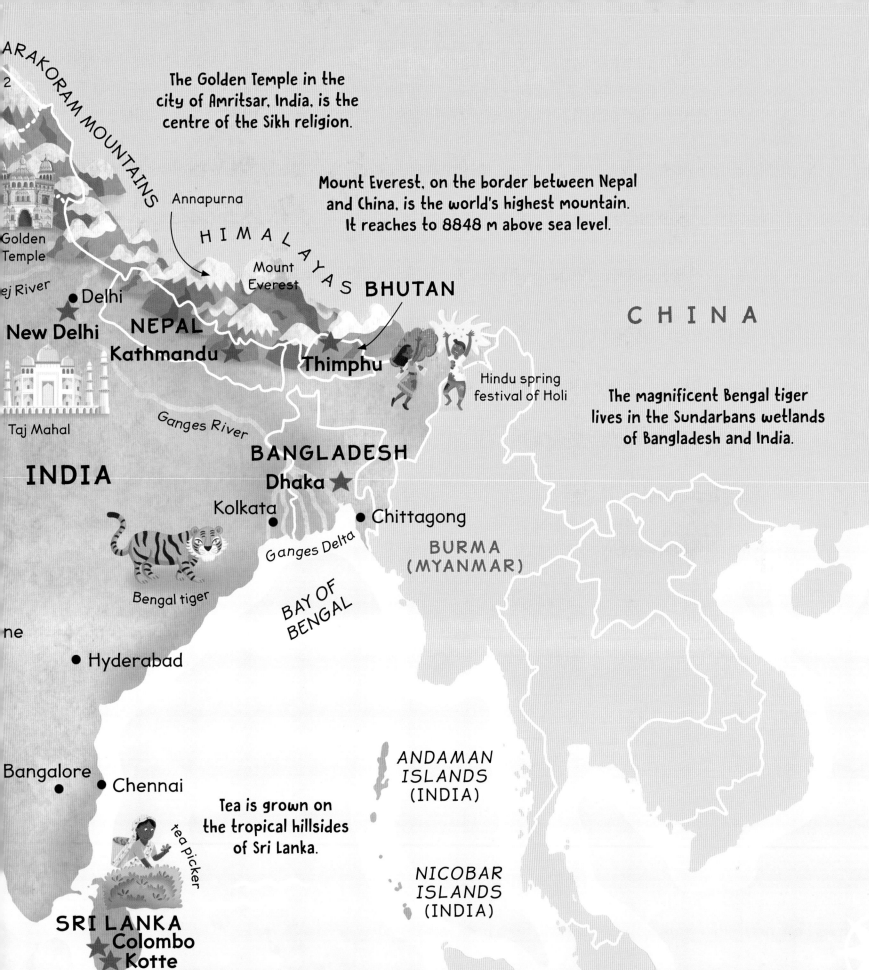

ARAKORAM MOUNTAINS

The Golden Temple in the city of Amritsar, India, is the centre of the Sikh religion.

Mount Everest, on the border between Nepal and China, is the world's highest mountain. It reaches to 8848 m above sea level.

Golden Temple

Annapurna

HIMALAYAS

BHUTAN

CHINA

ej River

Mount Everest

Delhi

NEPAL

New Delhi

Kathmandu

Thimphu

Hindu spring festival of Holi

The magnificent Bengal tiger lives in the Sundarbans wetlands of Bangladesh and India.

Taj Mahal

Ganges River

BANGLADESH

INDIA

Dhaka

Kolkata

Chittagong

Ganges Delta

BURMA (MYANMAR)

Bengal tiger

ne

Hyderabad

BAY OF BENGAL

ANDAMAN ISLANDS (INDIA)

Bangalore

Chennai

Tea is grown on the tropical hillsides of Sri Lanka.

tea picker

NICOBAR ISLANDS (INDIA)

SRI LANKA
Colombo
Kotte

37

RUSSIA

ALTAI MOUNTAINS

In ancient times the Great Wall of China was built across the north to defend it from attack. Its main section is almost 8900 km long.

Mongolia is one of the world's least crowded countries. One third of the population lives in the capital, Ulan Bator.

Ulan Bator ★

MONGOLIA

KAZAKHSTAN

GOBI DESERT

KYRGYZSTAN

TIAN SHAN

TAKLAMAKAN DESERT

Great Wall of China

Huang He (Yellow River)

TAJIKISTAN

KUNLUN SHAN

yak

CHINA

giant panda

AFGHANISTAN

K2

The giant panda lives in the bamboo forests of southwest China.

KARAKORAM MOUNTAINS

TIBET

TIBETAN PLATEAU

Wu

Annapurna

Chengdu ●

PAKISTAN

HIMALAYAS

Salween River

Chongqing ●

Chang Jiang (Yangtze River)

● Lhasa

INDIA

NEPAL

Mount Everest

Ho Ko

Mekong River

Pearl River

Can you find on the map the names of the two longest rivers in China?

BURMA (MYANMAR)

VIETNAM

Guangzhou

Mao

LAOS

Hainan

38

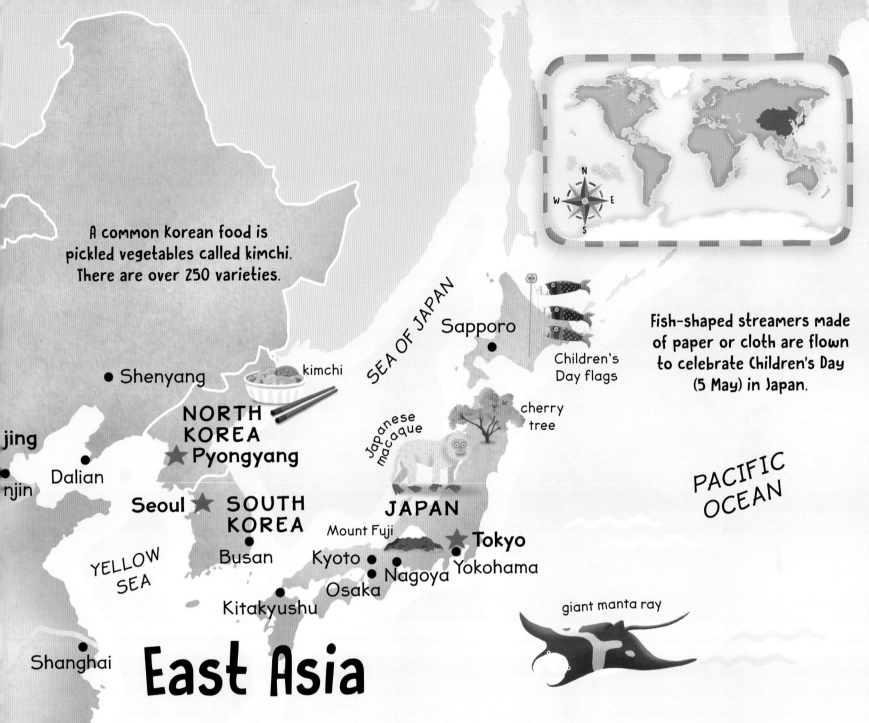

A common korean food is pickled vegetables called kimchi. There are over 250 varieties.

kimchi

SEA OF JAPAN

Sapporo

Children's Day flags

Fish-shaped streamers made of paper or cloth are flown to celebrate Children's Day (5 May) in Japan.

Shenyang

NORTH KOREA

★ Pyongyang

jing

Dalian

njin

Japanese macaque

cherry tree

Seoul ★

SOUTH KOREA

JAPAN

Mount Fuji

★ Tokyo

Busan

Kyoto

Yokohama

YELLOW SEA

Nagoya

Osaka

PACIFIC OCEAN

Kitakyushu

giant manta ray

Shanghai

# East Asia

Taipei

Taiwan

zhen

SOUTH CHINA SEA

Riders on horseback herd sheep and goats on the windy grasslands of Mongolia. In Western China the peaks of Tibet and the Himalayas form the 'roof of the world'. Millions of people live in huge cities in the east and south. Beyond North and South Korea are the islands of Japan, where high-tech cities contrast with peaceful mountains and rice fields.

INDIA

CHINA

BANGLADESH

*Irrawaddy River*

*Salween River*

1

Hanoi

Naypyidaw ★

Haiphong

3

rice field

Vientiane ★

*Mekong River*

2

Angkor Wat

BAY OF BENGAL

Yangon (Rangoon)

Bangkok ★

4    5

Phnom Penh ★

GULF OF THAILAND

The spectacular temples of Angkor Wat in Cambodia were built by the Khmer people in the 1100s.

Ho Chi Minh City

Thai stilt house

INDIAN OCEAN

Medan •

Ipoh •

Kuala Lumpur ← MALAYSIA

Putrajaya •

Bandar S Begaw

Johor Bahru

6 ★ Singapore

orang-u

Rafflesia flower

*Sumatra*

Rafflesia, from the jungles of Sumatra, is the world's biggest flower – and a real stinker!

Palembang •

Mount Merapi

IND

Jakarta ★

Surabo

*Java*

Bandung

N
W    E
S

40

# Southeast Asia

Two very long rivers flow across Southeast Asia. The Irrawaddy flows through the mountain valleys of Burma (Myanmar), while the Mekong crosses Burma, Laos, Thailand, Cambodia and Vietnam. These are tropical lands, with jungle villages, rice fields and water buffalo. Singapore, Kuala Lumpur and Jakarta are big centres of international business. Sumatra, Java, Borneo and countless other Indonesian islands stretch eastwards to the Philippines.

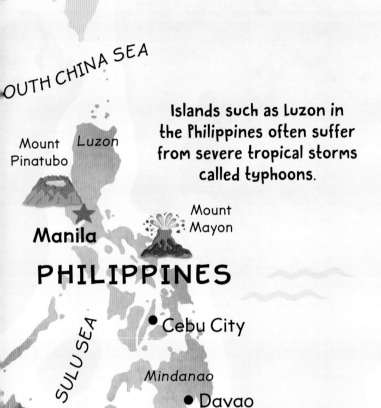

OUTH CHINA SEA

Mount Pinatubo

Luzon

**Manila**

## PHILIPPINES

Islands such as Luzon in the Philippines often suffer from severe tropical storms called typhoons.

Mount Mayon

SULU SEA

● Cebu City

Mindanao

● Davao

Mount Kinabalu

CELEBES SEA

neo

Sulawesi

SIA

Komodo dragon

mbawa

Flores

Timor

The Komodo dragon is the biggest lizard on Earth, measuring up to 3 m long.

★ **Dili**
← **EAST TIMOR**

New Guinea

PAPUA NEW GUINEA

Indonesia and the Philippines have many active volcanoes. Many of the lands around the Pacific Ocean are in an earthquake and volcano danger zone known as the 'Ring of Fire'.

# Australia and Papua New Guinea

Most people in Australia live on the coast, in large cities such as Sydney and Melbourne. Southern Australia is warm and sunny, but the northern coasts are tropical. Inland there are large farms for raising sheep, as well as vast areas of dry bush country and scorching deserts. Over 840 languages are spoken in Papua New Guinea. It is home to many different groups of people.

Queen Alexandra's birdwing butterfly

Queen Alexandra's birdwing is the world's biggest butterfly. It lives in the tropical forests of Papua New Guinea.

PACIFIC OCEAN

PAPUA NEW GUINEA

New Guinea

TORRES STRAIT

Port Moresby

GULF OF CARPENT

INDONESIA

dugong

Darwin

ilfish

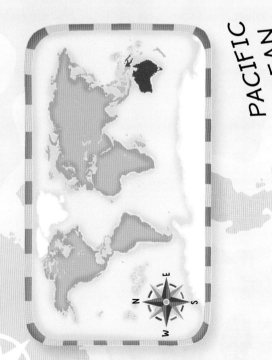

N
E
S
W

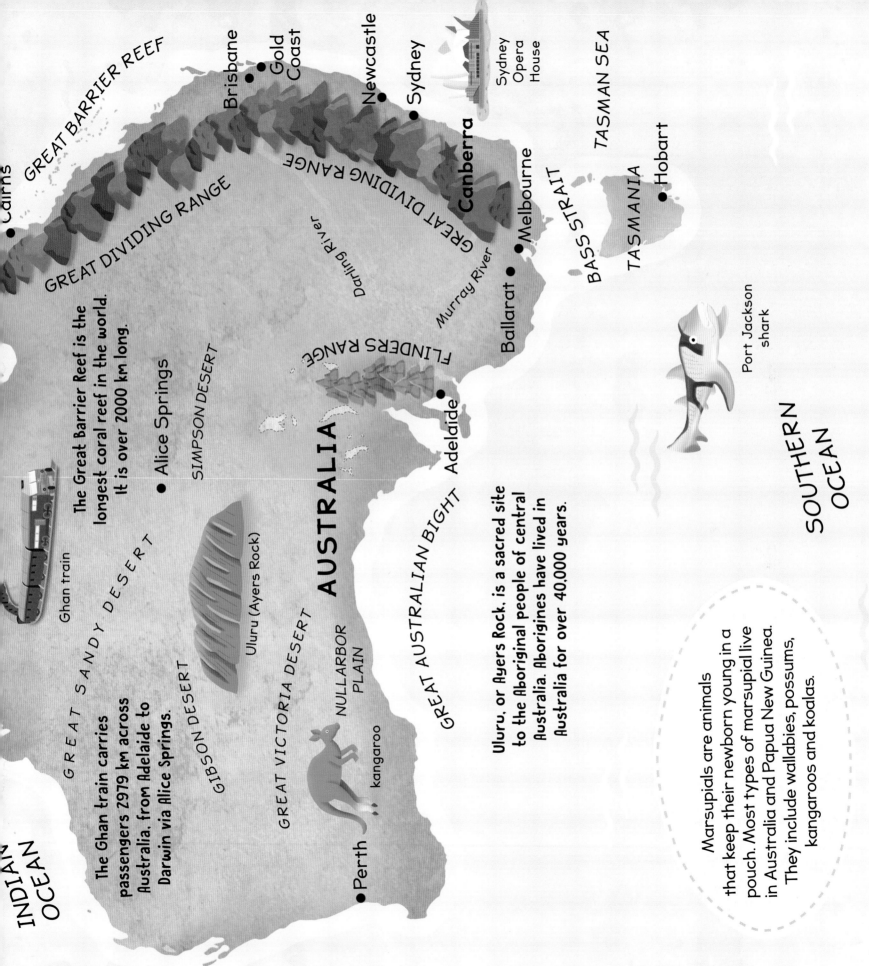

INDIAN OCEAN

GREAT BARRIER REEF

Cairns

GREAT DIVIDING RANGE

GREAT DIVIDING RANGE

Brisbane

Gold Coast

Newcastle

Sydney

Sydney Opera House

Canberra

Melbourne

Ballarat

Murray River

Darling River

FLINDERS RANGE

Adelaide

GREAT AUSTRALIAN BIGHT

AUSTRALIA

Alice Springs

SIMPSON DESERT

GREAT SANDY DESERT

Ghan train

GIBSON DESERT

Uluru (Ayers Rock)

GREAT VICTORIA DESERT

NULLARBOR PLAIN

kangaroo

Perth

TASMAN SEA

Hobart

BASS STRAIT

TASMANIA

Port Jackson shark

SOUTHERN OCEAN

The Great Barrier Reef is the longest coral reef in the world. It is over 2000 km long.

The Ghan train carries passengers 2979 km across Australia, from Adelaide to Darwin via Alice Springs.

Uluru, or Ayers Rock, is a sacred site to the Aboriginal people of central Australia. Aborigines have lived in Australia for over 40,000 years.

Marsupials are animals that keep their newborn young in a pouch. Most types of marsupial live in Australia and Papua New Guinea. They include wallabies, possums, kangaroos and koalas.

MARIANA TRENCH

spotted jellyfish

NORTHERN MARIANA ISLANDS (USA)

GUAM (USA)

MARSHALL ISLANDS

HAWAII (USA)
Honolulu

Islands in the Pacific are often formed from underwater volcanoes or coral reefs.

PALAU

FEDERATED STATES OF MICRONESIA

TOKELAU (NEW ZEALAND)

coconut palm tree

NAURU

seaplane

PAPUA NEW GUINEA

SOLOMON ISLANDS

KIRIBATI

TUVALU

★ Honiara

SAMOA
Apia ★

rugby player

VANUATU
Port Vila ★

NEW CALEDONIA (FRANCE)

Suva ★
FIJI

NIUE

COOK ISLANDS

AUSTRALIA

TONGA

WALLIS AND FUTUNA (FRANCE)

AMERICAN SAMOA (USA)

Sports teams from New Zealand often perform a haka before they compete. This ceremonial dance is a tradition of the Maori, the Polynesian people who were the first to settle in the islands, which they call Aotearoa.

North Island

NEW ZEALAND

Auckland ●

brown kiwi

South Island

SOUTHERN ALPS

★ Wellington

● Christchurch

Mount Cook

kiwis are forest birds that cannot fly. They are only found in New Zealand. People from New Zealand are often nicknamed 'kiwis'.

Red hot lava from Hawaii's Kilauea volcano flows into the sea.

auea
cano

NORTH PACIFIC OCEAN

# New Zealand and the Pacific islands

The Pacific Ocean covers about one third of Earth's surface and has many thousands of small islands. The main groups of islanders are known as Melanesians, Micronesians and Polynesians, but there are Europeans, Americans and Asians too. New Zealand is made up of two main islands. It has hot springs, snowy mountains and grasslands, where farmers raise sheep and cattle, and grow fruit.

coconut crab

FRENCH POLYNESIA (FRANCE)

eete

traditional Polynesian canoe

PITCAIRN ISLANDS (UK)

Easter Island statues

EASTER ISLAND (CHILE)

The Pacific Ocean is HUGE. The distance from Fiji to Tonga is 800 km.

Mysterious stone statues were created on Easter Island by the native Polynesian people, between 700 and 500 years ago.

SOUTH PACIFIC OCEAN

45

# The Arctic

In the centre of the Arctic Ocean is the northernmost point on our planet, the North Pole. Here the ocean is frozen all year round. Further from the Pole, the sea ice melts and breaks up during the short Arctic summer.

Each year, more and more sea ice is melting in the summer, because Earth is getting warmer. This change in the climate is already having a big effect on oceans, people, plants and animals.

The Arctic seas and ice floes are home to whales, walruses, fish, polar bears and seabirds.

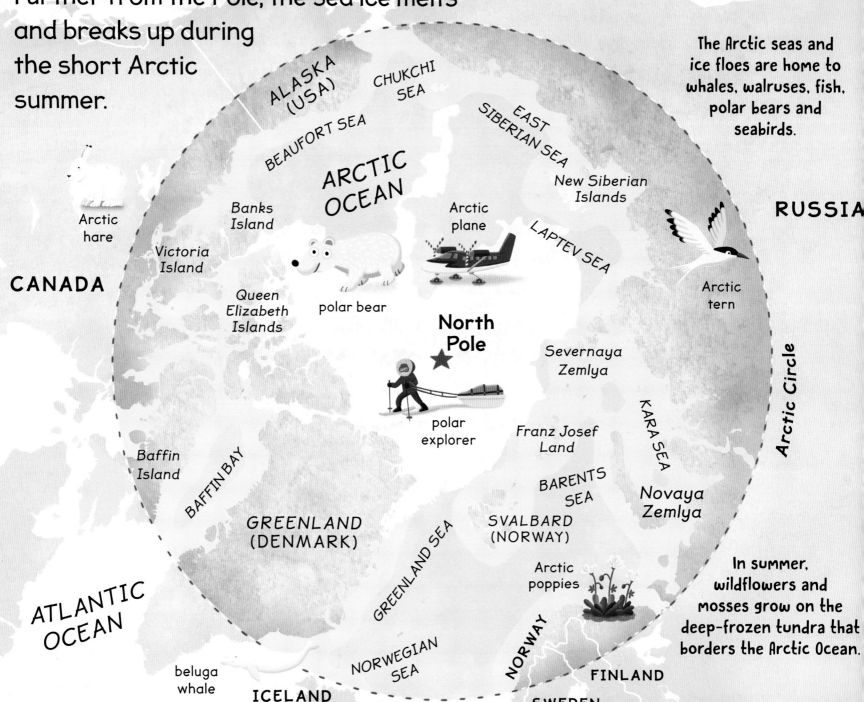

ALASKA (USA)

CHUKCHI SEA

BEAUFORT SEA

EAST SIBERIAN SEA

ARCTIC OCEAN

New Siberian Islands

Arctic plane

LAPTEV SEA

RUSSIA

Banks Island

Arctic hare

Victoria Island

CANADA

polar bear

Queen Elizabeth Islands

Arctic tern

North Pole

Severnaya Zemlya

polar explorer

Franz Josef Land

KARA SEA

Baffin Island

BAFFIN BAY

BARENTS SEA

Novaya Zemlya

GREENLAND (DENMARK)

GREENLAND SEA

SVALBARD (NORWAY)

Arctic Circle

ATLANTIC OCEAN

Arctic poppies

In summer, wildflowers and mosses grow on the deep-frozen tundra that borders the Arctic Ocean.

NORWAY

beluga whale

NORWEGIAN SEA

ICELAND

FINLAND

SWEDEN

46

# Antarctica

Antarctica is a vast desert covered in thick ice. It is the windiest place on Earth and bitterly cold. Sheets of ice stretch out to sea, where seals, penguins and whales swim in the freezing water. The South Pole is at the centre of Antarctica.

Mainland Antarctica was first seen in 1820. The first person to reach the South Pole was Norwegian explorer Roald Amundsen, in 1911.

It is too cold for people to live here permanently, but scientists come to study at special bases.

leopard seal

WEDDELL SEA

QUEEN MAUD LAND

ENDERBY LAND

Southern royal albatross

scientific station

Antarctic krill

Antarctic Peninsula

Ronne Ice Shelf

EAST ANTARCTICA

SOUTHERN OCEAN

ELLSWORTH MOUNTAINS

★ South Pole

PRINCESS ELIZABETH LAND

AMUNDSEN SEA

WEST ANTARCTICA

MARIE BYRD LAND

TRANSANTARCTIC MOUNTAINS

polar vehicle

WILKES LAND

orca

Ross Ice Shelf

iceberg

emperor penguins

ROSS SEA

Mount Erebus

GEORGE V LAND

Emperor penguins breed on the Antarctic ice shelves during the harsh winter.

Antarctic Circle

SOUTHERN OCEAN

47

# Index

First published in 2017 by Miles Kelly Publishing Ltd
Harding's Barn, Bardfield End Green,
Thaxted, Essex, CM6 3PX, UK

Copyright © Miles Kelly Publishing Ltd 2017

This edition printed 2021

10 12 14 15 13 11 9 7

**Publishing Director** Belinda Gallagher
**Creative Director** Jo Cowan
**Editorial Director** Rosie Neave
**Illustration/design** Alistar Illustration (including cover) and Emanuela Carletti from Milan Illustrations Agency, Rob Hale, Simon Lee, Joe Jones, Stephan Davis
**Senior Editor** Amy Johnson
**Production** Jennifer Brunwin
**Image Manager** Liberty Newton
**Reprographics** Stephan Davis
**Assets** Lorraine King

ISBN 978-1-78617-224-2

Printed in China

British Library Cataloguing-in-Publication Data
A catalogue record for this book is available from the British Library

Made with paper from a sustainable forest

www.mileskelly.net